BATS

Rebecca Woodbury, Ph.D., M.Ed.

Gravitas Publications Inc.

Bats

Illustrations: Janet Moneymaker

Bats
ISBN 978-1-953542-31-1

Published by Gravitas Publications Inc.
Imprint: Real Science-4-Kids
www.gravitaspublications.com
www.realscience4kids.com

RS4K

Photo credits: Cover & Title Page: By Natalia, AdobeStock; Above: NPS Photo by Shawn Thomas; P. 5, Nils Bouillard on Unsplash; P. 9. By Ishan @seefromthesky on Unsplash; P. 11. By Nobbi, CC BY SA 3.0; P. 13. Serrah Galos on Unsplash; P. 15. Zdeněk Macháček on Unsplash; P. 17. By chamnan phanthong, AdobeStock; P. 19. Susanne Martinus on Unsplash; P. 20. vishu vishuma on Unsplash

Have you ever spent time
outside when it is getting dark?
Have you seen what looks like
birds zipping around in the sky?

You may be seeing **bats!**

What a pretty face.

Bats are a type of **mammal.**

I am a type of
mammal too.

Review: MAMMALS

Mammals are animals that have some things in common. They:

- Breathe air.

- Feed milk to their babies.

- Have fur for at least part of their life.

- Most mammals give birth to live babies.

Bats are the only
flying mammal.

How about flying squirrels?

They don't really fly. They just glide.

Bats sleep during the day in caves, trees, and other protected places. They hang upside down while they are sleeping.

I'm so happy I don't have to sleep like this!

At night, bats fly around
to catch bugs to eat.

Not all bats eat bugs.
Some eat fruit and plant
parts and may drink
nectar from flowers.

What a long tongue!

It can reach the nectar inside the flowers.

Baby bats are called **pups**. Most pups hold on to their mother while she is flying around to get food.

Pups? Like dogs?

Yes.

Bats can live a long time.

Some live to be over 30 years old.

There is still much to learn about bats and why we need them!

Some Bat FACTS!

More about bats:

- Bats help control insect populations.

- They pollinate plants and spread seeds.

- Most have only one pup per year.

- Many gather in large groups at night.

- They are not related to mice.

- Bats won't get tangled in your hair.

- They won't suck your blood.

- Bats are not blind.

How to say science words

bat   (BAT)

insect   (IN-sekt)

mammal   (MAA-muhl)

nectar   (NEK-tuhr)

pollinate   (PAH-luh-nayt)

population   (pah-pyuh-LAY-shun)

pup   (PUHP)

science   (SIY-uhns)

www.ingramcontent.com/pod-product-compliance
Lightning Source LLC
Chambersburg PA
CBHW040149200326
41520CB00028B/7537